ONE
BLUE
GNU

by Danna Smith · illustrated by Ana Zurita

For Teresa Costanza, a friend I can count on—D.S.

To Ana and Javi and all the parties we will do together—A.Z.

Text copyright © 2022 Danna Smith
Illustrations copyright © 2022 Ana Zurita

Published in 2022 by Amicus Ink, an imprint of Amicus
P.O. Box 227 • Mankato, MN 56002
www.amicuspublishing.us

Library of Congress Cataloging-in-Publication Data
Names: Smith, Danna, author. | Zurita, Ana, illustrator.
Title: One blue gnu / by Danna Smith; illustrated by Ana Zurita.
Description: Mankato, Minnesota : Amicus Ink, [2022] | Audience: Ages 4-8 | Summary: "When
   a box of cell phones is accidentally delivered to the zoo, one blue gnu quickly calls two
   white sheep, who plan a party - beep beep beep! Call by call, the party spreads around
   the zoo, but oh no! Who invited tiger, and who will he call on his new phone? A fun and
   colorful picture book romp, where we count from one to ten and down again in one
   huge party!"—Provided by publisher.
Identifiers: LCCN 2020012943 (print) | LCCN 2020012944 (ebook) |
   ISBN 9781681527451 (hardcover) | ISBN 9781681527468 (ebook)
Subjects: LCSH: Zoo animals—Juvenile literature. | Counting—Juvenile
   literature. | Gnus—Juvenile literature.
Classification: LCC QL77.5 S575 2021 (print) | LCC QL77.5 (ebook) |
   DDC 590—dc23
LC record available at https://lccn.loc.gov/2020012943
LC ebook record available at https://lccn.loc.gov/2020012944

Editor: Rebecca Glaser • Designer: Kathleen Petelinsek

First Edition 9 8 7 6 5 4 3 2 1
Printed in China

ONE
BLUE
GNU

by Danna Smith · illustrated by Ana Zurita

amicus ink

Mankato, Minnesota

One blue gnu, home all alone.
One blue gnu with a new cell phone.

Making calls to two white sheep,
she plans a party—beep beep beep.

Three orange apes on a tire swing
answer their new phones when they ring.

Four red pandas in the sun
agree a party would be fun.

4

Five green ducks out on the lake
laugh at the sound the cell phone makes.

Six yellow snakes receive a call.
The laughing ducks invite them all.

Seven black yaks share a tasty snack.
They miss Snake's call,
but they call right back.

Eight pink pigs who like to snooze
receive a message Meet at Gnu's 🙂.

⑧

Sleepy pigs awake from slumber.
They call the hippos—
"Oops, wrong number!"

Nine gray hippos in the shade
hear of the plan that Blue Gnu made.

10

Ten purple birds hear a cell phone chime.
They fly away; it's PARTY TIME!

Fifty-four friends soon arrive.
Plus, one blue gnu
makes fifty-five.

GREAT SNACK FOR SEVEN BLACK YAKS

TODAY'S BIRTHDAY

ZOO NEWS

PARTY AT GNU'S

Fifty-five beasts just keeping cool,
splashing around in Blue Gnu's pool.

Pigs and pandas learn to float.

Yaks speed by in a blow-up boat.

Animals dive and birds keep score.

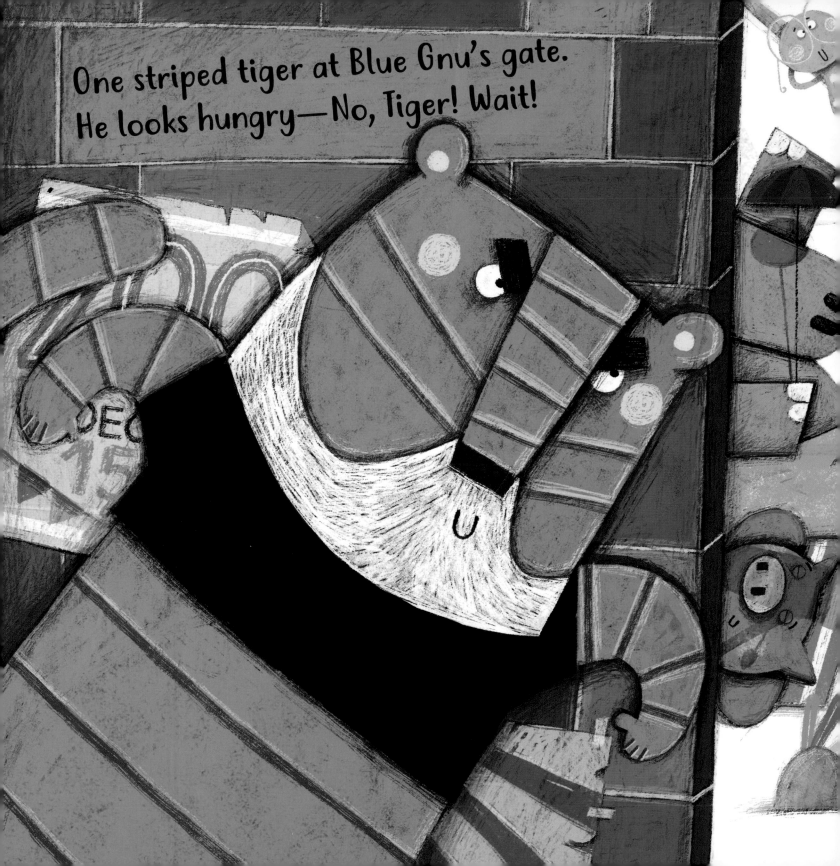

One striped tiger at Blue Gnu's gate.
He looks hungry—No, Tiger! Wait!

Creeping close, he has a plan...

The tiger calls a pizza man!

Fifty-six friends eat their lunch—
a wet and wild and beastly bunch.

Blue Gnu's friends are fed and dry,
the time has come to say goodbye.

10 Ten purple birds take off in flight.

Nine gray hippos slip from sight. **9**

Eight pink pigs hug their friend Gnu.
Seven black yaks bid her adieu.

Six yellow snakes slither back home.
Five green ducks and Tiger roam.

Four red pandas hitch a ride.
Three orange apes leave side by side.

Two white sheep wish they could stay.

Now all Gnu's friends have gone away.

One blue gnu, home all alone,
receives a message on her phone.
She swipes the screen and takes a peek...

Big party at our pen next week!

1